D1825236

DIVE IN!

EXPLORING THE OCEAN ZONES

Jessica Domer

BeaLu Books

ISBN Hardcover: 978-1-7341065-1-0
ISBN Paperback: 978-1-7333092-4-0

Library of Congress Control Number: 2019952460
Publisher's Cataloging-in-Publication Data is on file with the publisher.

Edited by: Luana K. Mitten
Book cover and interior design by Tara Raymo • creativelytara.com

Printed in the United States of America
October 2019

BeaLu Books
Tampa, Florida

www.BeaLuBooks.com

PHOTO CREDITS: Cover: © wildestanimal,© Jane Rix, © Gerald Robert Fischer; Page 1: © Rich Carey; Page 4-5: © jannoon028, © robuart, © dive-hive; Page 6: © Lee Yiu Tung, © andrea crisante; Page 7: © Romolo Ravani, © divedog; Page 8-9: © Rich Carey, © Lorna Roberts, © Wanida_Sri, © janicsteohart, © Naoto Shinkai, © megablaster; Page 10: © sumroeng chinnapan, © noaa, © Choksawatdikorn, © Willyam Bradberry, © Watchares Hansawek; Page 11: © pixinoo; Page 12-13: © creativesunday, © nazz lpez, © vojce, © Ethan Daniels, © Yilmaz Tuncel; Pages 14-15: ©creativesunday, © noaa, © Luiz Felipe V Puntel, © Vladimer Wrangel; Page 16: © Vladi333

TABLE OF CONTENTS

THE OCEAN

The sunlight sparkles on the surface of the ocean water. Don't you want to dive in and play? I know I do! Let's dive in together and explore!

SAFETY FIRST!

Swimming in the ocean can be dangerous! Swim where it is marked safe for swimming and have a grown-up with you.

OCEANS AROUND THE WORLD

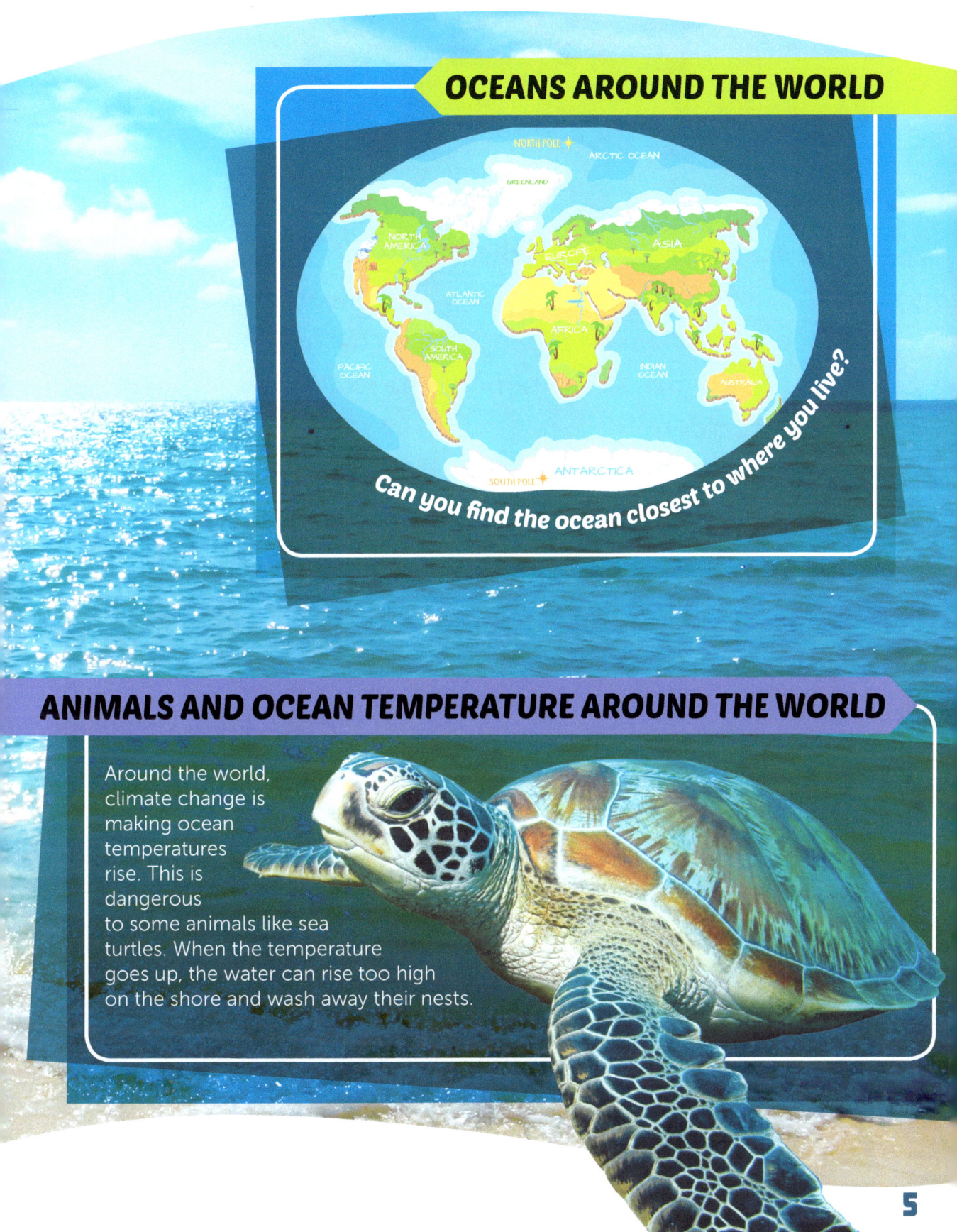

NORTH POLE ★

ARCTIC OCEAN

GREENLAND

NORTH AMERICA

EUROPE

ASIA

ATLANTIC OCEAN

PACIFIC OCEAN

SOUTH AMERICA

AFRICA

INDIAN OCEAN

AUSTRALIA

SOUTH POLE ★

ANTARCTICA

Can you find the ocean closest to where you live?

ANIMALS AND OCEAN TEMPERATURE AROUND THE WORLD

Around the world, climate change is making ocean temperatures rise. This is dangerous to some animals like sea turtles. When the temperature goes up, the water can rise too high on the shore and wash away their nests.

OCEAN ZONES

Before we get started, let's plan our journey to the bottom of the ocean! The ocean is divided into three main zones that go all the way from the very surface of the water to the deep, deep bottom. At the top, the water gets more sunlight making it warmer. At the bottom, the water gets less sunlight making it colder.

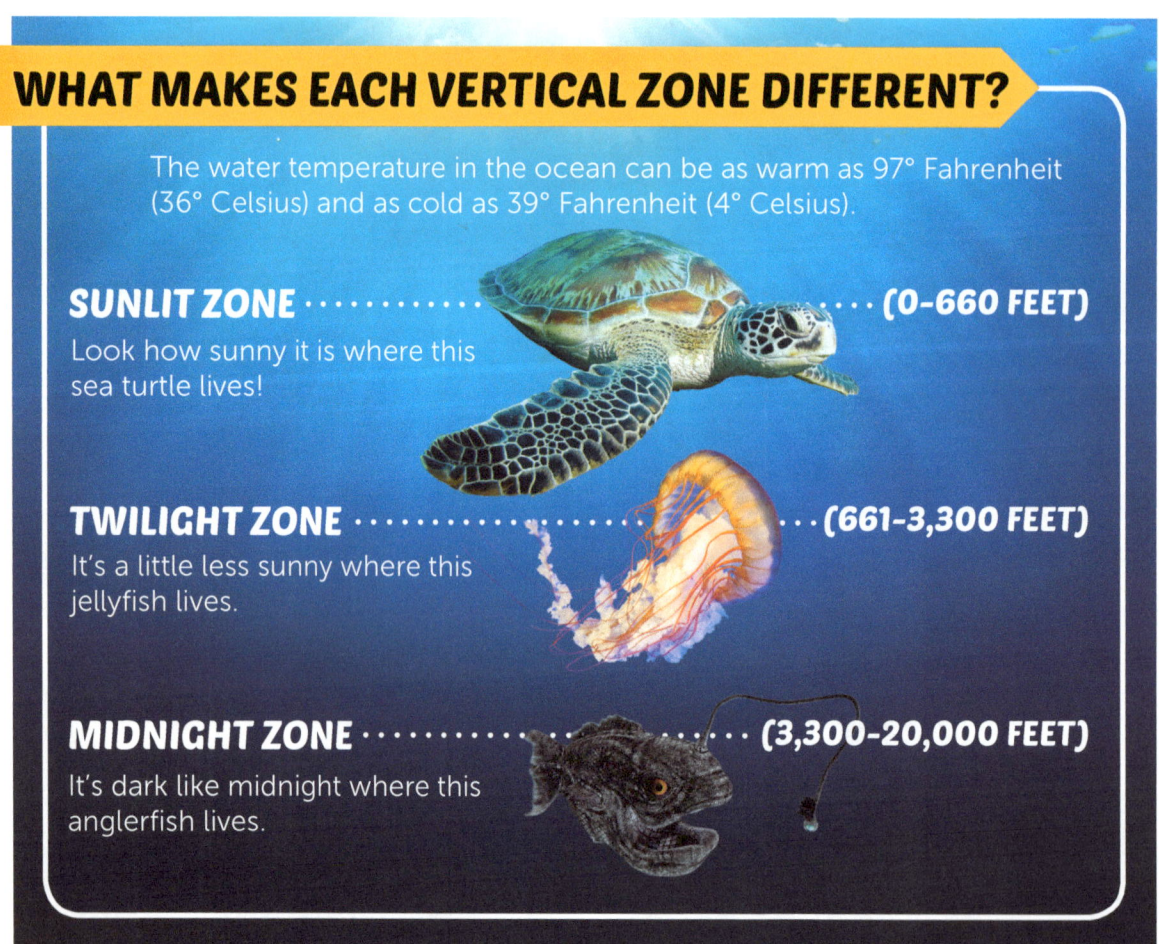

WHAT MAKES EACH VERTICAL ZONE DIFFERENT?

The water temperature in the ocean can be as warm as 97° Fahrenheit (36° Celsius) and as cold as 39° Fahrenheit (4° Celsius).

SUNLIT ZONE · **(0-660 FEET)**
Look how sunny it is where this sea turtle lives!

TWILIGHT ZONE · · · · · · · · · · · · · · · · · · **(661-3,300 FEET)**
It's a little less sunny where this jellyfish lives.

MIDNIGHT ZONE · · · · · · · · · · · · · · **(3,300-20,000 FEET)**
It's dark like midnight where this anglerfish lives.

THE SUNLIT ZONE: CLOSE TO THE SURFACE

Oh, the sunny sunlit zone! My favorite place. That's because there's so much warm sunlight. Most of the ocean's animals live here because there are more food sources for them. There is enough sunlight in this zone for plants to grow, too. Another name for the sunlit zone is the euphotic zone.

THE LIGHT OF LIFE: PHOTOSYNTHESIS

Photosynthesis is the process that plants use to grow by changing sunlight to energy. Seaweed uses photosynthesis to grow, just like a plant that grows in the ground!

ANIMALS OF THE SUNLIT ZONE

Sea turtles, sharks, sea lions, stingrays, and lots of other animals make waves in this zone. Let's take a moment to meet just a few. Some you might already know, others you might get to meet for the first time!

The bluefin tuna lives in the sunlit zone of the Indian, Pacific, and Southern Atlantic oceans. It is endangered because of overfishing. The bluefin tuna is often served in sushi, a popular Japanese dish.

The orange-and-white clownfish and the anemone have a symbiotic relationship, which means they have adapted to help each other. The anemone kills its prey by injecting them with poison. The clownfish is immune to the anemone's poison. This dynamic duo works because the clownfish's bright colors get the attention of prey for the anemone, and then the clownfish gets to eat the leftovers.

The humpback whale is a big kid. Although it grows to be around 52 feet long, it is gentle and playful, and it is known for the beautiful songs it sings to communicate with other humpbacks. It eats small fish, so it is endangered because of the overfishing of its food sources. Humpback whales can also get stuck in fishing nets or be hit by boats.

The weedy sea dragon and leafy sea dragon are masters of disguise. They live in the Atlantic Ocean in the Australian Kelp Forest, and they have adapted by using camouflage to protect themselves from predators. Both have some of the most beautiful costumes of all the ocean's creatures.

CAMOUFLAGED ABOVE, CAMOUFLAGED BELOW

Some sea creatures use a type of camouflage known as countershading. Their bellies are light in color, like the sunlight above them. Their backs are darker in color, like the murky water below them. This lets them hide from predators looking down at them from above or up at them from below.

leopard seal

THE SUNLIT ZONE'S ECOSYSTEM:
FROM THE SUN TO THE TABLE

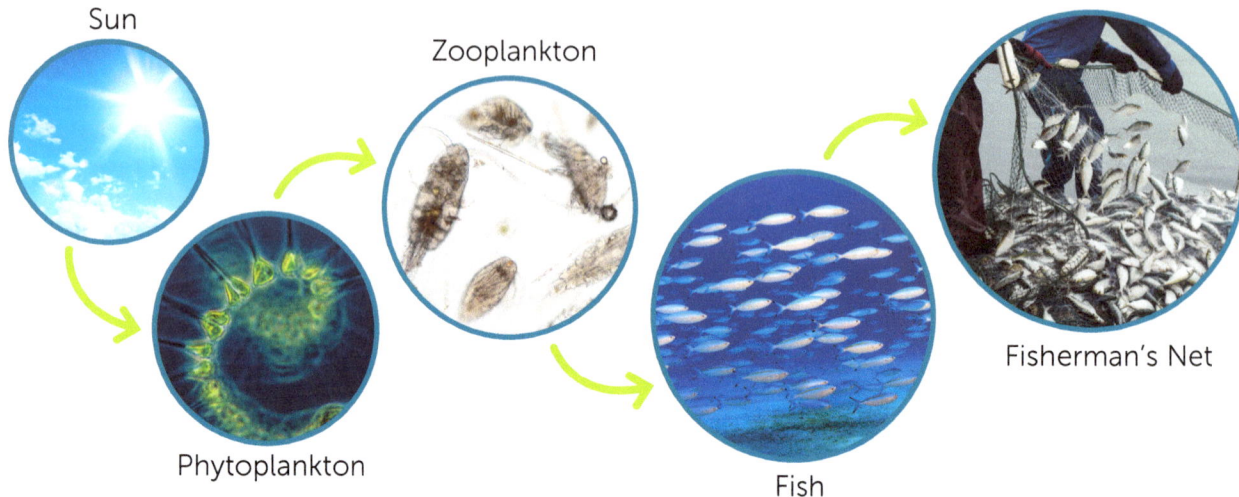

Sun

Zooplankton

Phytoplankton

Fish

Fisherman's Net

Life in the sunlit zone is connected like the threads of a fisherman's net. From the twinkling sun to the teeny plankton to the tremendous whale, each part of the sunlight zone plays an important role in the food chain. Microscopic phytoplankton are food for small ocean creatures called zooplankton. Large fish eat the zooplankton and are caught in the fisherman's net for us to eat. How lucky we are to be part of the sunlit zone's ecosystem!

OVERFISHING AND ITS CONSEQUENCES

Overfishing happens when fishermen catch boatful after boatful of the fish we like to buy at the store and eat in restaurants. When they catch too many fish, it leaves not enough fish for other members of the ocean's ecosystem, which can harm some animals.

RESPONSIBLE SHOPPERS CAN HELP THE OCEAN!

You can help stop overfishing by buying fish to eat that are not endangered due to overfishing. Check online for resources to find out more!

THE TWILIGHT ZONE: IN THE MIDDLE

The twilight zone, or disphotic zone, has less light than the sunlit zone, so there are fewer plants and animals in this zone. The animals that do live here are clever. They have adapted to survive the darker colder water by having bigger eyes and smaller bodies. Being small helps them hide from predators.

Changing ocean temperatures caused by global climate change can melt polar ice caps, making the twilight zone brighter than it should be. Researchers believe this could be a problem for creatures of the twilight zone.

BIOLUMINESCENCE

Bioluminescence allows animals in the twilight zone to glow in the dark water like fireflies glow at night. Scientists have even found that the animals are brighter than the light from the sun and stars in the twilight zone!

comb jellies

ANIMALS OF THE TWILIGHT ZONE

Starfish live in the twilight zone, but they are getting a name change to sea stars because they are not actually fish. They are echinoderms, a type of ocean animal. These creatures eat using their feet to suction food from the ocean floor. Their stomachs come out of their mouths on their bellies and take in the food. Sea stars have no brains or blood! Their circulatory system pumps water, not blood.

This lantern fish lives in the twilight zone. It uses bioluminescence to shine in the dark. It has another special adaptation, too. It has large eyes that help it see in the dark.

Dead or alive, sand dollars are cool creatures that live in the twilight zone. The hard, white sand dollars you find washed up on the beach are only the exoskeleton of dead sand dollars. A living sand dollar is covered with thousands of purple spines.

THE MIDNIGHT ZONE: THE NEW FRONTIER

The mysterious midnight zone, also known as the aphotic zone, is teaching scientists more about life on planet Earth. The midnight zone is the deepest, darkest, and coldest zone. It is a completely dark part of the ocean. Life here is hard—but animals are able to survive in unique ways. Scientists are discovering new types of ecosystems that do not use photosynthesis to create life. Instead, they use a process called chemosynthesis.

Chemosynthesis happens because cracks in the ocean's floor called hydrothermal vents produce hot, toxic gases. Bacteria feed on these gases and produce energy that is used by creatures in the midnight zone. Another source of food in the midnight zone is food particles that fall down from the higher ocean zones.

AN IMPORTANT DISCOVERY: HYDROTHERMAL VENTS

Before hydrothermal vents and their ecosystems were discovered, scientists believed that the only process that could make energy for living things was photosynthesis.

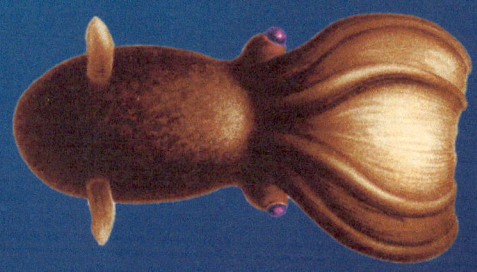

Vampire squid only look scary. They are named for their huge eyes and webbed arms that look like a vampire's cloak. These creatures of the midnight zone aren't even predators! They use two long sticky tentacles to collect drifting particles called marine snow for their meals.

Survival in the twilight zone is tough. To give themselves a fighting chance, basket stars have developed the ability to regrow body parts when they are lost. These creatures have superpowers!

The chimaera fish has a unique way of finding prey without the light of day. It has tiny dots on its face that can sense electricity, which lets it find moving prey to eat. It has no bones, so it is not crushed by the pressure of all the water at the depths of the ocean.

KEEP DIVING IN
PROTECTING OUR OCEANS FOR THE FUTURE

Oceans are Earth's blanket, watering can, and refrigerator. They cover two-thirds of the Earth's surface, provide the water that keeps our planet hydrated through the water cycle, and supply an important and healthy source of food. Because of environmental threats to our oceans, such as climate change, overfishing, and pollution it is important to learn more about how to protect our sparkling blue treasures — our oceans.

WHAT CAN YOU DO?

- Ways to keep plastics from getting in the oceans:
    - Buy clothes made of natural fibers like bamboo and hemp
    - Carry a reusable water bottle
    - Put purchases in reusable shopping bags
    - Decorate for celebrations without using balloons (and never release balloons outside)
    - Skip the straw in your drink

- Ways to reduce global warming by reducing your carbon footprint:
    - Go into a fast-food restaurant instead of using the drive-through
    - Walk, bike, or carpool to where you need to go
    - Turn off or unplug electronic devices when not in use
    - Buy food and products that are grown or made locally

ABOUT THE AUTHOR

Jessica Domer loves teaching kids about language because interesting words make reading and writing more fun! She has a really nice cat named Lucky, who is friendly and playful. Lucky also walks all over the laptop while Ms. Domer writes! See if you can find any cat hairs in your book.

Read more!

https://oceanservice.noaa.gov/kids/

www.montereybayaquarium.org/conservation-and-science

https://climatekids.nasa.gov

www.amnh.org/explore/ology/marine-biology

CPSIA information can be obtained
at www.ICGtesting.com
Printed in the USA
LVHW071711041219
639416LV00001B/2/P